HURRICANES VS. TORNADOES VS. TYPHOONS:
WIND SYSTEMS OF THE WORLD

SPEEDY PUBLISHING

Speedy Publishing LLC
40 E. Main St. #1156
Newark, DE 19711
www.speedypublishing.com

HURRICANES, TORNADOES
AND TYPHOONS ARE BOTH
STORMY ATMOSPHERIC SYSTEMS
THAT HAVE THE POTENTIAL
TO CAUSE DESTRUCTION.

HURRICANES

A hurricane is a tropical cyclone, occurring in the North Atlantic Ocean or the Northeast/North-Central Pacific Ocean. Hurricanes develop over warm water and use it as an energy source.

Hurricanes can be up to 600 miles wide. They move slowly over the ocean, gaining power and speed. Hurricane winds can blow up to 200 miles per hour.

Hurricanes have led to the death of around 2 million people over the last 200 years.

Hurricanes are classified into 5 categories, based on their wind speeds and potential to cause damage.

TORNADOES

A tornado is a violently rotating column of air that is in contact with both the surface of the earth and a cumulonimbus cloud.

Most tornadoes have wind speeds less than 110 miles per hour. Extreme tornadoes can reach wind speeds of over 300 miles per hour. Tornadoes are sometimes called twisters.

Tornados form on every continent in the world except for Antarctica, but the most common place for them is in the United States.

There are over 1,000 tornadoes in the U.S. every year.

TYPHOONS

A typhoon is a mature tropical cyclone that develops in the western part of the North Pacific Ocean.

Nearly one-third of the world's tropical cyclones form within the western Pacific. Pacific typhoons have formed year round, with peak months from August to October.

They can last more than a week. When they hit land, they can cause flooding and destroy buildings and cars.

Printed in Great Britain
by Amazon

37797211R00020